Mapographica

the

MANMADE WORLD

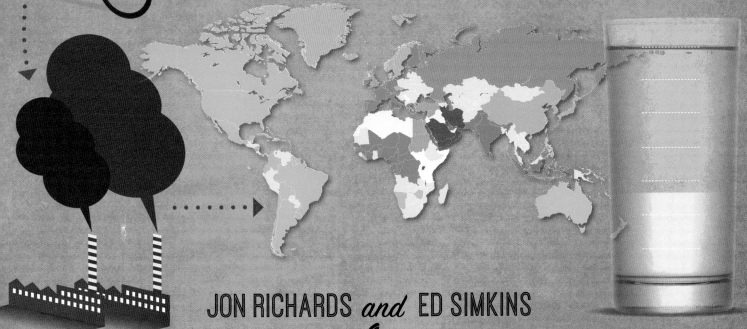

JON RICHARDS *and* ED SIMKINS

WAYLAND

CONTENTS

ACKNOWLEDGEMENTS

First published in Great Britain in 2015
by Wayland
Copyright © Wayland, 2015
All rights reserved
Editor: Julia Adams
Produced for Wayland by Tall Tree Ltd
Designer: Ed Simkins
Editor: Jon Richards

Dewey number: 304.2'0223-dc23
ISBN 978 0 7502 9145 3

Wayland, an imprint of
Hachette Children's Group
Part of Hodder and Stoughton
Carmelite House
50 Victoria Embankment
London EC4Y 0DZ

An Hachette UK Company
www.hachette.co.uk
www.hachettechildrens.co.uk

Printed and bound in Malaysia

10 9 8 7 6 5 4 3 2 1

Picture credits can be found on page 32

— The world in —
100 PEOPLE

If you reduced the world's population to just 100 people, then an 'average' person would not own a computer or use the internet. However, they would own a mobile phone and have access to electricity and safe water, and they would be living on more than US$2 a day.

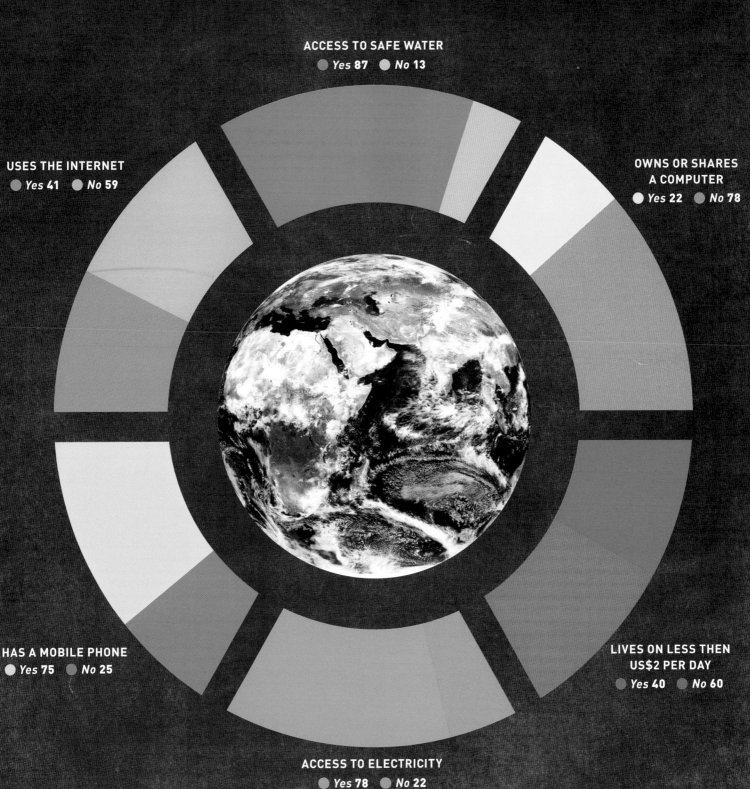

ACCESS TO SAFE WATER
● *Yes 87* ● *No 13*

USES THE INTERNET
● *Yes 41* ● *No 59*

OWNS OR SHARES A COMPUTER
● *Yes 22* ● *No 78*

HAS A MOBILE PHONE
● *Yes 75* ● *No 25*

LIVES ON LESS THEN US$2 PER DAY
● *Yes 40* ● *No 60*

ACCESS TO ELECTRICITY
● *Yes 78* ● *No 22*

People and ENERGY

This image shows the world at night, revealing where the main concentrations of people live. However, the brightest areas aren't always the most populated, as richer countries will use more electricity per person, and appear brighter, than poorer ones.

THE WORLD AT NIGHT

Toronto

Canada has a population of just 3.5 people per square kilometre.

Each person in the USA uses 12,500 kWh.

CANADA
Area: 10 million square kilometres
Population: 35 million

Most Canadians live in a few large cities, such as Toronto, leaving vast areas with very few people that are very dark in this image.

USA
Energy consumption:
4 trillion kilowatt hours (kWh)
Population: 318.9 million

The USA has a population of about 320 million. However, the people in this rich country use so much energy that they are the second largest users of electricity in the world.

THE WORLD'S POPULATION USES MORE THAN **20 TRILLION KWH** OF ELECTRICITY A YEAR.

The USA accounts for about 20 per cent of this.

BIGGEST ENERGY CONSUMERS
(BILLION KWH)

Russia
1,037

India
757.9

China
4,831

USA
3,883

Japan
859.7

BIGGEST POPULATIONS

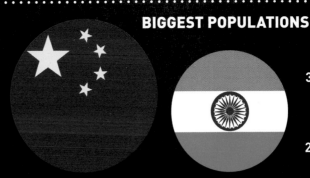

China
1.335 billion

India
1.236 billion

USA
318.9 million

Indonesia
253.6 million

Brazil
202.7 million

EGYPT/THE NILE
Nearly all of Egypt's population of about 87 million people live in a thin band on either side of the Nile.

Two Koreas
The Korean peninsula is split into two nations: North Korea and South Korea. The tiny dot of light in the northern half shows the location of North Korea's capital city – Pyongyang.

NORTH KOREA
Energy consumption: 18 billion kWh
Population: 25 million

Pyongyang

SOUTH KOREA
Energy consumption: 450 billion kWh
Population: 50 million

NIGERIA
Energy consumption:
20.4 billion kWh
Population: 177 million

The population of Nigeria is more than half that of the USA. However, each person only uses 115 kWh – less than one tenth of the amount used by a citizen of the USA. This is why the country appears so dark.

Each person in Nigeria uses 115 kWh.

Brisbane

Perth

Sydney

Melbourne

AUSTRALIA
Area: 7.7 million square kilometres
Population: 22.5 million

Nearly 55 per cent of Australia's population of 22.5 million live in its four largest cities – Sydney, Melbourne, Brisbane and Perth – which lie dotted along the country's east and western coasts.

A Moving WORLD

The way people choose to travel varies greatly from one country to another. While some nations have huge rail networks, populations of other countries prefer to travel by car or even bicycle.

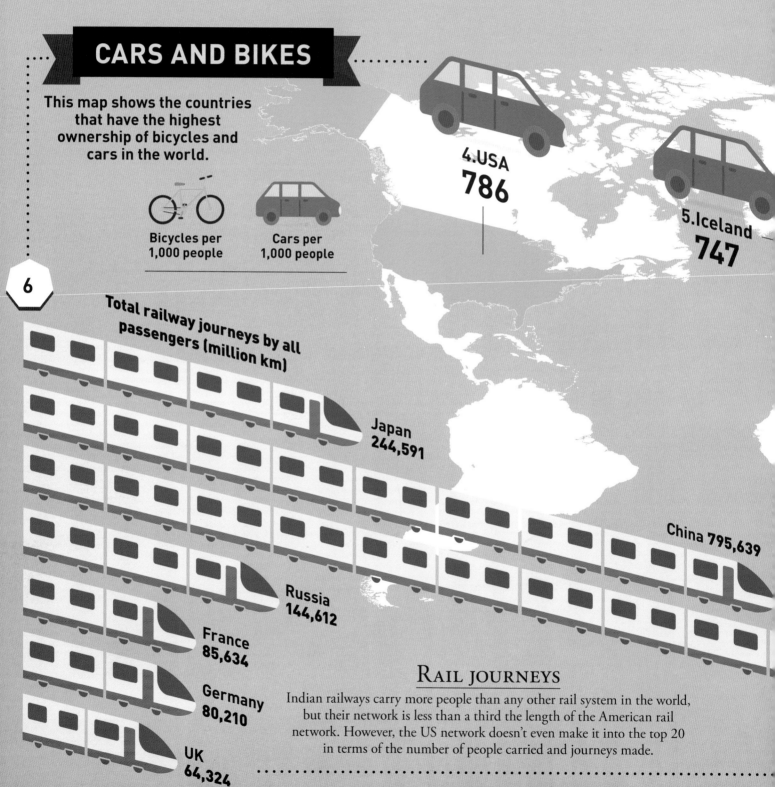

CARS AND BIKES

This map shows the countries that have the highest ownership of bicycles and cars in the world.

Bicycles per 1,000 people

Cars per 1,000 people

4.USA
786

5.Iceland
747

Total railway journeys by all passengers (million km)

Japan
244,591

China **795,639**

Russia
144,612

France
85,634

Germany
80,210

UK
64,324

RAIL JOURNEYS

Indian railways carry more people than any other rail system in the world, but their network is less than a third the length of the American rail network. However, the US network doesn't even make it into the top 20 in terms of the number of people carried and journeys made.

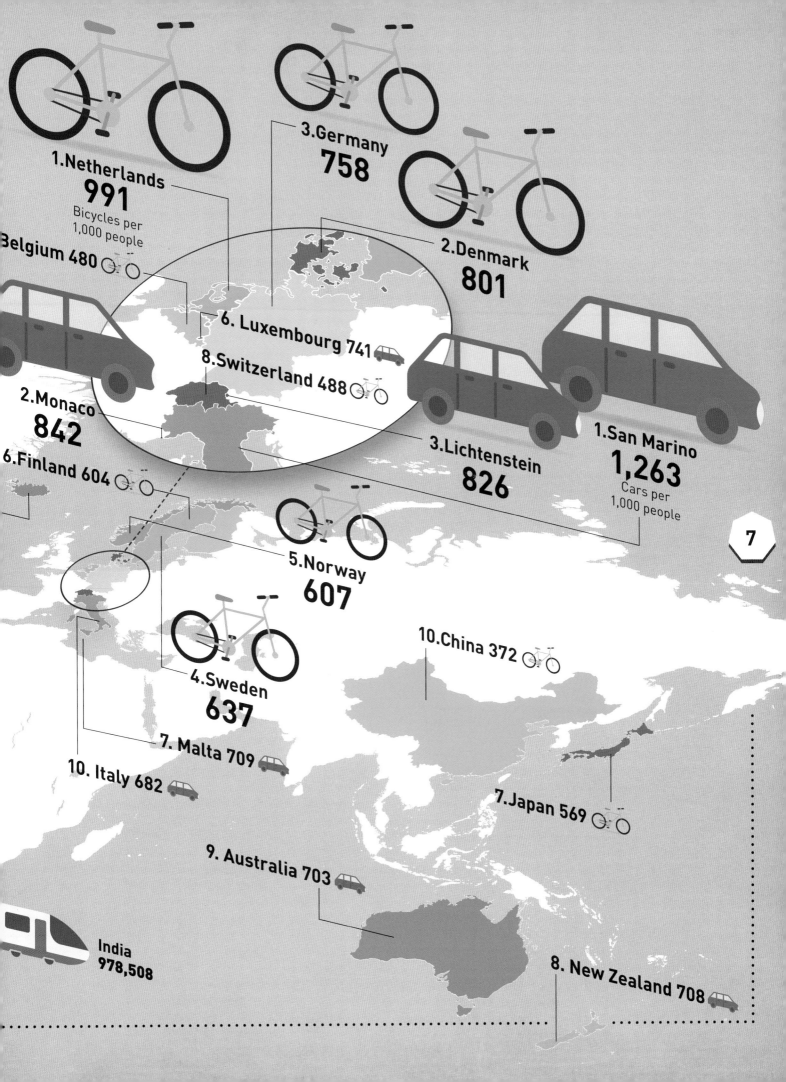

1.Netherlands
991
Bicycles per
1,000 people

3.Germany
758

2.Denmark
801

Belgium 480

6. Luxembourg 741

8.Switzerland 488

2.Monaco
842

6.Finland 604

3.Lichtenstein
826

1.San Marino
1,263
Cars per
1,000 people

7

5.Norway
607

10.China 372

4.Sweden
637

7. Malta 709

10. Italy 682

7.Japan 569

9. Australia 703

India
978,508

8. New Zealand 708

TRADE

Which goods countries trade and ship to other nations depends on their accessibility to raw materials. Countries that are short in raw materials may choose to import them (ship them in) from countries that export them (ship them out) and use them to manufacture other goods.

HIGHEST VALUE EXPORT

This map shows the major type of export for each country in terms of the amount of money it earns.

TYPE OF EXPORT

- OIL AND GAS
- FOOD AND DRINK
- METALS AND MINERALS
- PRECIOUS METALS AND MINERALS
- TEXTILES AND CLOTHES
- MACHINERY AND TRANSPORTATION
- ELECTRONICS
- WOOD
- OTHER

Fish

Motor Vehicles and Parts

Capital Goods

Electrical Equipment
Machinery
Motor Vehicles
Fish
Diamonds
Manufactured Goods
Computers
Machinery
Engineering Products
Agricultural Products
Machinery
Clothing
Fo

Clothes and shoes
Petroleum
Petroleum
Sugar
Aluminium
Transport Equipment
Phosphates
Uranium
Cotton
Fish
Natural Gas
Iron
Diamonds
Rubber
Cotton
Cocoa Beans
Timber
Oil
Petroleum
Diamonds
Oil

Petroleum
Petroleum
Copper
Natural Gas
Transport Equipment
Soybeans
Beef
Soybeans

Diamonds
Diamonds
Gold

IMPORT-EXPORT

China is the world's largest exporter because it has a large source of affordable labour and has developed a huge manufacturing industry. The USA is the world's largest importer because it is the wealthiest country in the world and its population demands goods that are made around the globe.

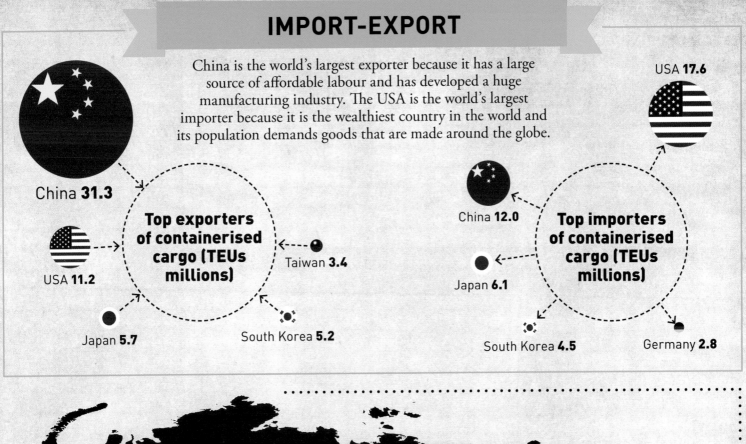

China **31.3**

USA **11.2**

Japan **5.7**

Top exporters of containerised cargo (TEUs millions)

Taiwan **3.4**

South Korea **5.2**

USA **17.6**

China **12.0**

Japan **6.1**

Top importers of containerised cargo (TEUs millions)

South Korea **4.5**

Germany **2.8**

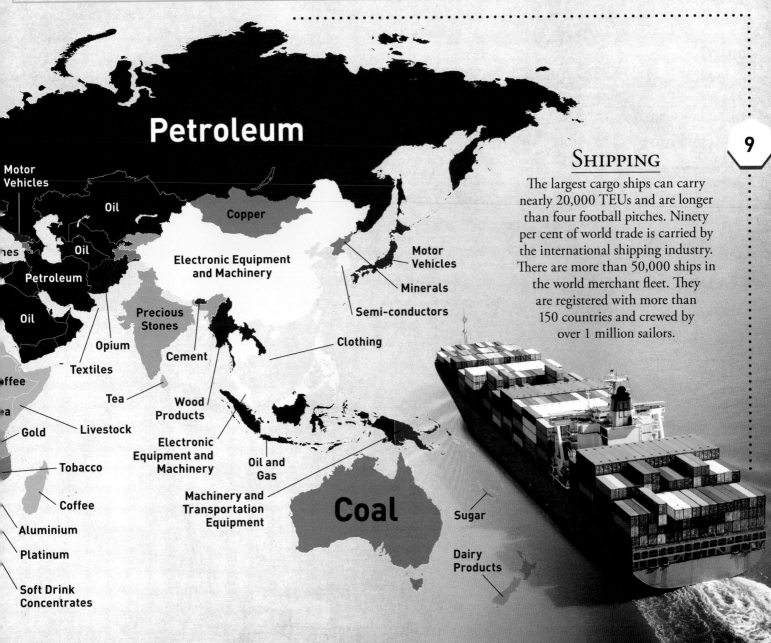

Petroleum

Motor Vehicles

Oil

Copper

Electronic Equipment and Machinery

Motor Vehicles

Minerals

Semi-conductors

nes

Oil

Petroleum

Oil

Precious Stones

Opium

Cement

Clothing

Textiles

Tea

Wood Products

offee

Gold

Livestock

a

Tobacco

Electronic Equipment and Machinery

Oil and Gas

Coffee

Aluminium

Platinum

Soft Drink Concentrates

Machinery and Transportation Equipment

Coal

Sugar

Dairy Products

9

SHIPPING

The largest cargo ships can carry nearly 20,000 TEUs and are longer than four football pitches. Ninety per cent of world trade is carried by the international shipping industry. There are more than 50,000 ships in the world merchant fleet. They are registered with more than 150 countries and crewed by over 1 million sailors.

Palm Oil
PRODUCTION

The fruit of the oil palm produces an oil that is used in a wide range of products – in fact, it can be found in nearly half of everything you buy. Today, nearly 60 million tonnes of palm oil are produced each year and the surge in demand for this substance in the last 50 years has led to a dramatic increase in deforestation rates in countries growing the crop.

PALM OIL PRODUCTION

This map shows the increase in annual production rates in millions of tonnes by the world's major palm oil producers between 1964 and 2014.

THAILAND
2014 **2.25**
1964 **0**

DEMOCRATIC REPUBLIC OF THE CONGO
2014 **0.21**
1964 **0.13**

NIGERIA
2014 **0.93**
1964 **0.54**

COLOMBIA
2014 **1.02**
1964 **0**

HONDURAS
2014 **0.44**
1964 **0**

ECUADOR
2014 **0.57**
1964 **0**

WHAT IS IT?

The oil palm is a tree that can grow 20 metres tall. It produces fruit that is processed to produce the oil at a rate of about 4.5 tonnes of oil per hectare each year. The leftover fibre is used to make animal feed.

chocolate

pizza dough

ice cream

margarine

Each piece of fruit contains
50% OIL

biodiesel

detergent

shampoo

soap

lipstick

USES FOR PALM OIL

Palm oil is the most widely used vegetable oil on the planet. It is used to make foodstuffs, such as pizza, ice cream and chocolate. It is also found in cleaning materials, including shampoo and detergent, as well as cosmetics and soap. It is even used to produce biodiesel fuel.

DEFORESTATION

Every year, millions of hectares of forest are cleared to make room for palm oil plantations. Between 2000 and 2012, Thailand lost about 6 million hectares of forest, which is about half the size of England, for plantations. As the forest disappears, so animal species, such as the orang-utan, face threats to their existence.

IN THE TIME IT WILL TAKE YOU TO READ THIS PAGE 2 FOOTBALL PITCHES OF **RAINFOREST** WILL HAVE BEEN CLEARED TO MAKE WAY FOR PALM OIL PLANTATIONS.

MALAYSIA

INDONESIA

2014 **20.35**
1964 **0.15**

2014 **33.5**
1964 **0.16**

OIL

Every day, 89.08 million of barrels of oil are pumped out of the ground and carried along pipelines and in enormous tankers to refineries. Here, the oil is processed to produce petroleum and other products, including gas and plastics.

BIGGEST OIL PRODUCERS

This map shows the world's biggest producers of oil. These countries pump oil out of the ground using land-based wells or huge sea-based oil rigs.

Canada
3.9
million barrels per day

USA
11.12
million barrels per day

Mexico
2.9
million barrels per day

WHAT OIL IS USED FOR?

Crude oil is transported to large oil refineries, where it is treated and turned into a wide range of products. The image below shows some of the major products created by refining oil. In the USA, petrol accounts for 46 per cent of refined oil products, while heating oil and diesel account for 20 per cent.

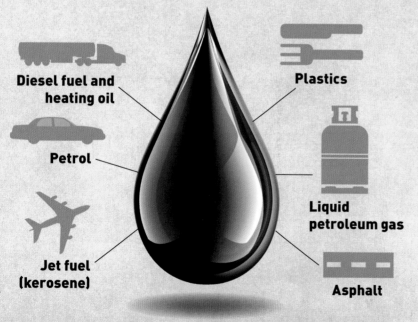

Diesel fuel and heating oil

Petrol

Jet fuel (kerosene)

Plastics

Liquid petroleum gas

Asphalt

BIGGEST IMPORTERS

Some countries have such high energy demands that they need to import oil as well as producing it. The USA, for example, is one of the world's biggest oil producers, but it's also the world's biggest importer of oil.

Top world oil net importers (thousand barrels per day)

= 1,000 barrels

Germany 2,225

China 5,608

India 2,460

France 1,699

USA 7,372

Japan 4,559

South Korea 2,261

Spain 1,272

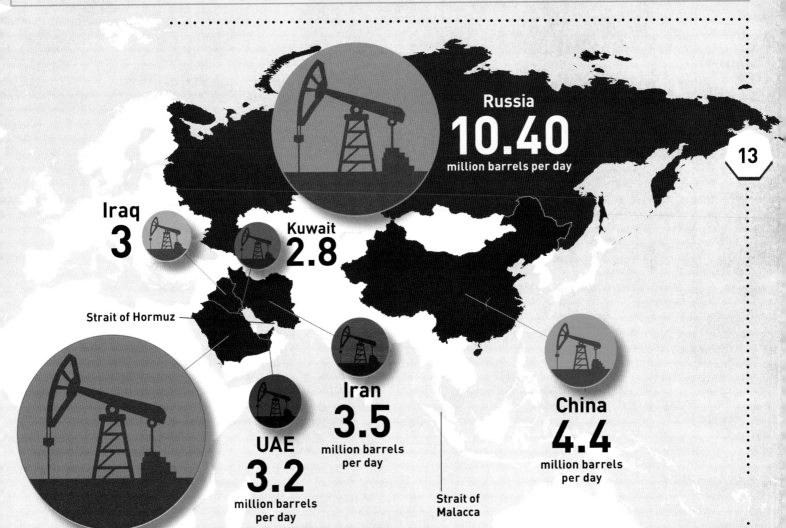

Russia
10.40
million barrels per day

Iraq
3

Kuwait
2.8

Strait of Hormuz

Saudi Arabia
11.73
million barrels per day

UAE
3.2
million barrels per day

Iran
3.5
million barrels per day

Strait of Malacca

China
4.4
million barrels per day

13

BUSIEST SHIPPING ROUTES

There are a number of points where sea routes are narrow, creating busy shipping lanes. The world's busiest shipping routes for oil are the Strait of Hormuz (17 million barrels per day) and the Strait of Malacca (15.2 million barrels per day).

Water ACCESS

Water is vital to humans, and we use 4,000 cubic km of it ever year – that's more water than in Lake Huron, USA. We drink it to stay alive, it irrigates our crops and it is used in all types of industry. However, not everyone on the planet has access to clean, safe water.

FRESHWATER ACCESS

While some countries have access to plenty of water, others have problems making sure that all of their people get enough water. This map shows the levels of water that are naturally present in each country and the countries with the largest populations who cannot get water that is safe to drink and free from contamination.

SUDAN
18 MILLION

ETHIOPIA
46 MILLION PEOPLE

NIGERIA
66 MILLION PEOPLE

DR CONGO
36 MILLION PEOPLE

TANZANIA
21 MILLION

KENYA
17 MILLION

WATER SUPPLY

ABSOLUTE SCARCITY

CHRONIC SCARCITY

STRESS

RELATIVE SUFFICIENCY

SANITATION

A reliable way of removing and treating human waste is essential to provide safe drinking water and prevent the spread of disease. People living in countries without these services are prone to diseases such as cholera, diarrhoea and typhoid.

2.5 BILLION PEOPLE AROUND THE WORLD DO NOT HAVE ACCESS TO ADEQUATE SANITATION FACILITIES.

CHINA
119 MILLION PEOPLE

NGLADESH
28 MILLION

INDONESIA
43 MILLION PEOPLE

INDIA
97 MILLION PEOPLE

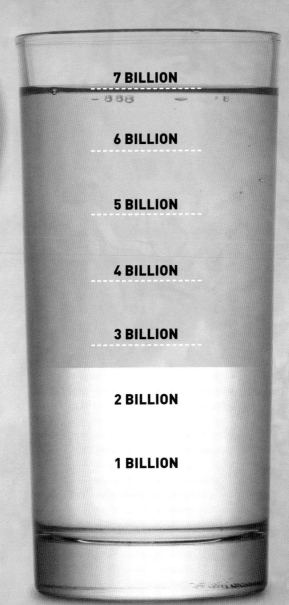

7 BILLION

6 BILLION

5 BILLION

4 BILLION

3 BILLION

2 BILLION

1 BILLION

15

THAT'S OVER 35% OF THE WORLD

Growing
FOOD

In order to feed its population, a country can grow the food on farms, or import food from countries that produce an excess. However, some countries are too poor to buy all the food they need.

IMPORTERS AND EXPORTERS

This map shows the world's biggest importers and exporters of food produce in billions of US dollars per year. It also shows the countries with the largest proportion of undernourished people.

USA
$55.7 BN

UK
$36.3 B

FRANCE
$45.3 BN

Imports

Exports

Haiti

USA
$72.6 BN

COUNTRIES WITH HIGHEST LEVELS OF UNDERNOURISHMENT
(PERCENTAGE OF POPULATION UNDERNOURISHED)

HAITI **52%**
ZAMBIA **48%**
CENTRAL AFRICAN REPUBLIC **38%**
NORTH KOREA **38%**
NAMIBIA **37%**

KEY
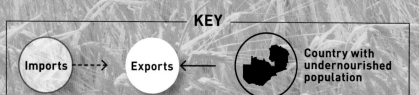

Imports - - - ▶   Exports ◀—   Country with undernourished population

Agricultural land is not distributed evenly across countries. In general, high-income countries have more cultivated land per person:

The UN's Food and Agriculture Organisation (FAO) states that 4.4 billion hectares of land could be used for crops. That's nearly 35 per cent of the world's total land area of about 13 billion hectares. We currently use just 1.6 billion hectares.

Medium-income country
0.23 hectares per person

Low-income country
0.17 hectares per person

High-income country
0.37 hectares per person

35%
4.4 billion hectares

BELGIUM
$29.2 BN

NETHERLANDS
$47 BN

CHINA
$43.3 BN

North Korea

GERMANY
$52 BN

GERMANY
$38.5 BN

JAPAN
$41.6 BN

Central African Republic

Zambia

Namibia

1960

2010

The World's agricultural productivity increased by 150–200% between 1960 and 2010.

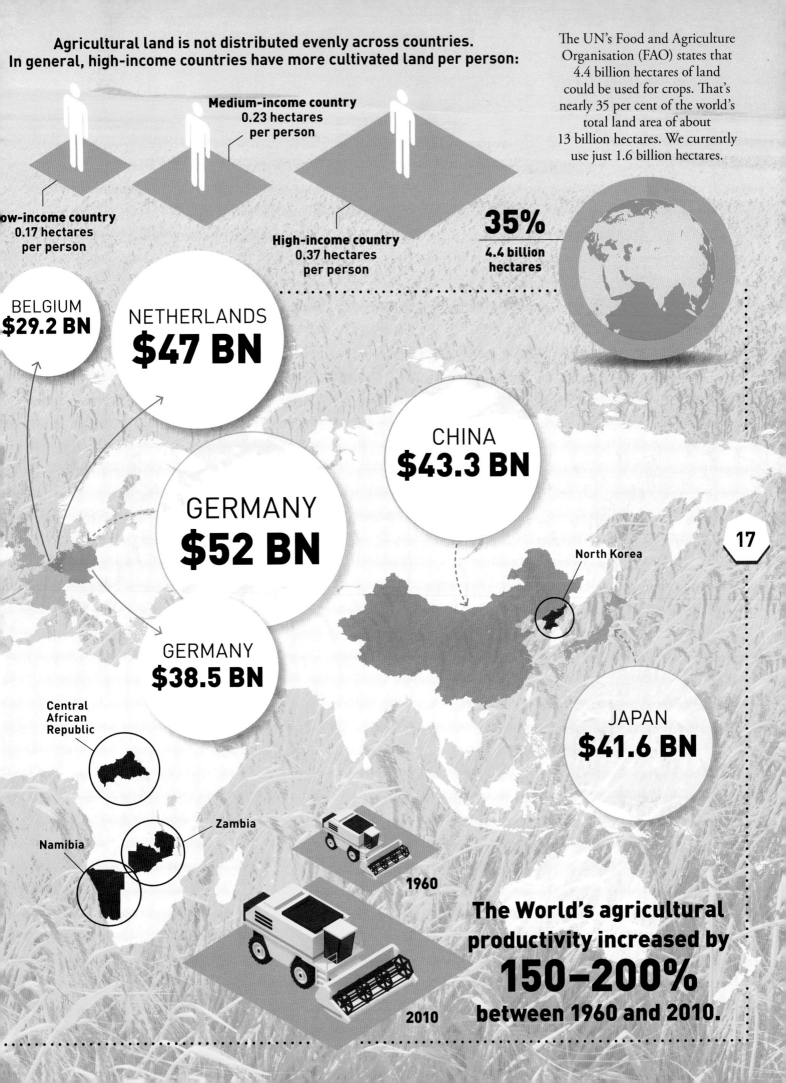

Global emissions and POPULATION

Carbon dioxide (CO_2) is the most common greenhouse gas that is responsible for climate change. It is present naturally and it is released by human activities such as energy production and industry.

WORLD'S BIGGEST CO_2 EMITTERS

This world map shows the countries with the largest populations and those that release the most CO_2. It also shows those countries whose people produce the most CO_2 per head.

2.USA
5,300,000 kt

3.USA 318.9 million

2.Trinidad and Tobago 37.2

5.Brazil 202.7 million

5.Aruba 23.9

KEY

World's biggest polluters greatest CO_2 emissions in kilotonnes (Kt)

These countries are the highest emitters of CO_2 because they have the largest populations.

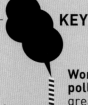

World's biggest populations

Highest CO_2 emissions per person (tonnes per person)

The most polluting nations (per person) are much smaller nations whose main industry is linked to petrochemicals.

Since 1751, **about 1,480 gigatonnes (billions of tonnes) of carbon dioxide** have been released by industrial activity.

More than half of this **(50.2% or 743 gigatonnes)** has been released since 1988.

HISTORICAL EMISSIONS

The countries listed here have released the greatest combined amounts of carbon dioxide since 1850.

USA 361,300.0 million tonnes

China 140,860.3 million tonnes

Russia 101,116.7 million tonnes

Germany 84,123.6 million tonnes

UK 70,042.3 million tonnes

1.China
10,330,000 kt

4.Russia
1,800,000 kt

5.Japan
1,360,000 kt

1.China
1.335 billion

3.Kuwait
29.1

3.India
2,070,000 kt

1.Qatar
43.9

4.Indonesia
253.6 million

The world produces about 35,270,000 kt of CO$_2$ every year.

2.India
1.236 billion

4.Brunei
Darussalam
24.0

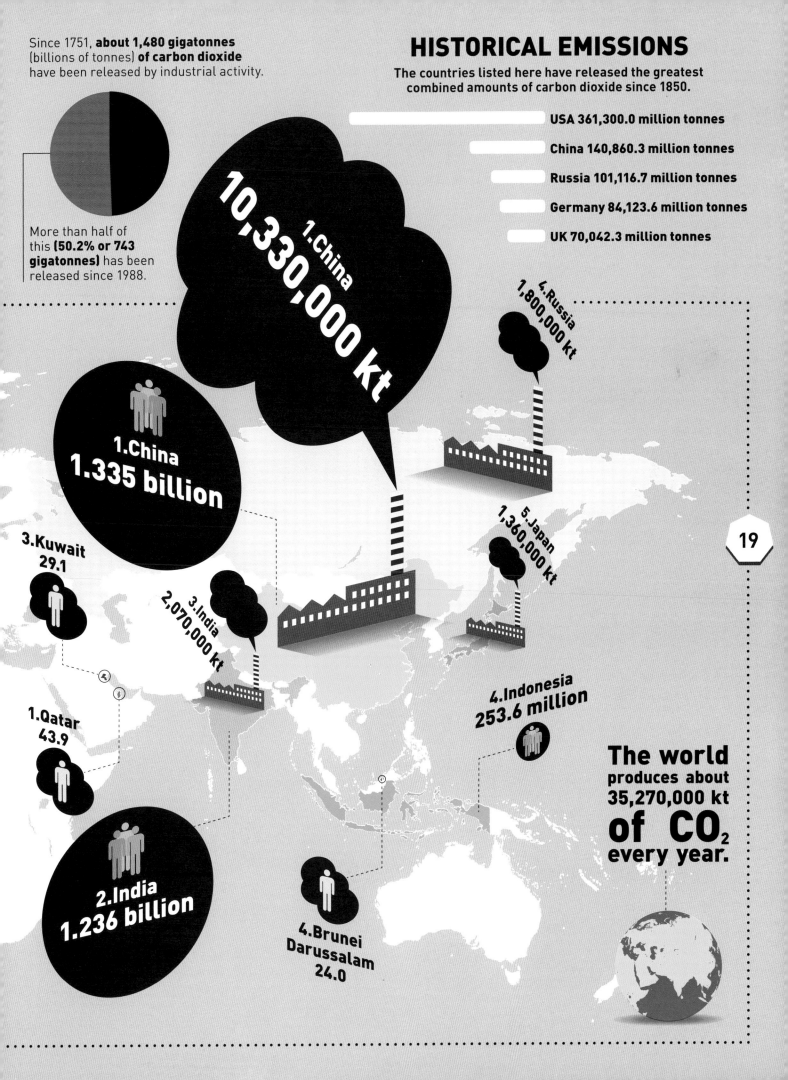

Tallest BUILDINGS

With space in many cities being limited, architects and planners choose to build up rather than spread out, creating ever taller skyscrapers with dozens of floors. Today's giants are nearly a kilometre tall and contain more than 100 storeys.

TALLEST BUILDINGS ON EACH CONTINENT

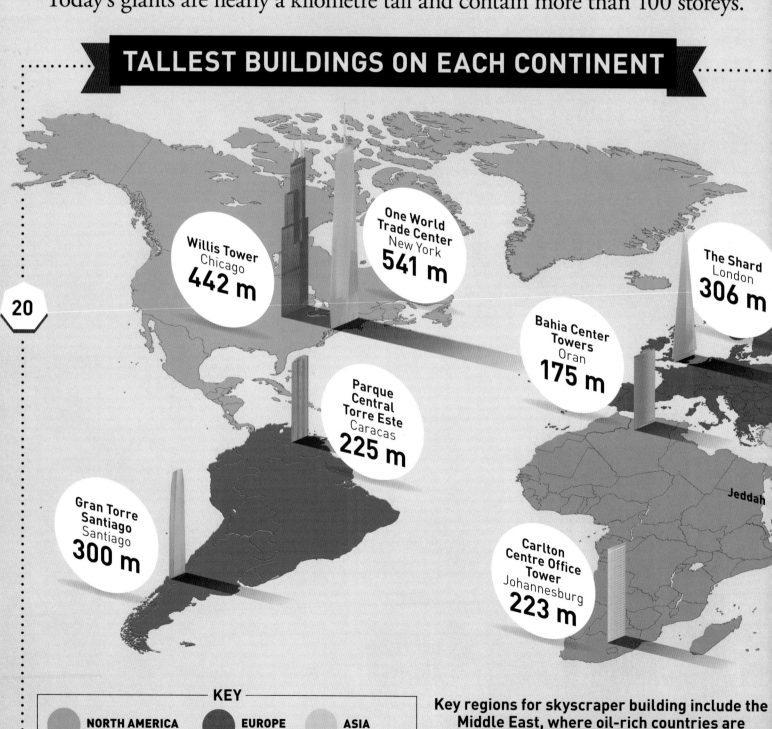

Willis Tower Chicago **442 m**

One World Trade Center New York **541 m**

The Shard London **306 m**

Bahia Center Towers Oran **175 m**

Parque Central Torre Este Caracas **225 m**

Gran Torre Santiago Santiago **300 m**

Carlton Centre Office Tower Johannesburg **223 m**

Jeddah

KEY
- NORTH AMERICA
- EUROPE
- ASIA
- SOUTH AMERICA
- AFRICA
- OCEANIA

Key regions for skyscraper building include the Middle East, where oil-rich countries are investing their wealth in huge building plans, and Russia, whose capital city, Moscow, has seven of Europe's ten tallest buildings. They were all built in the last 10 years.

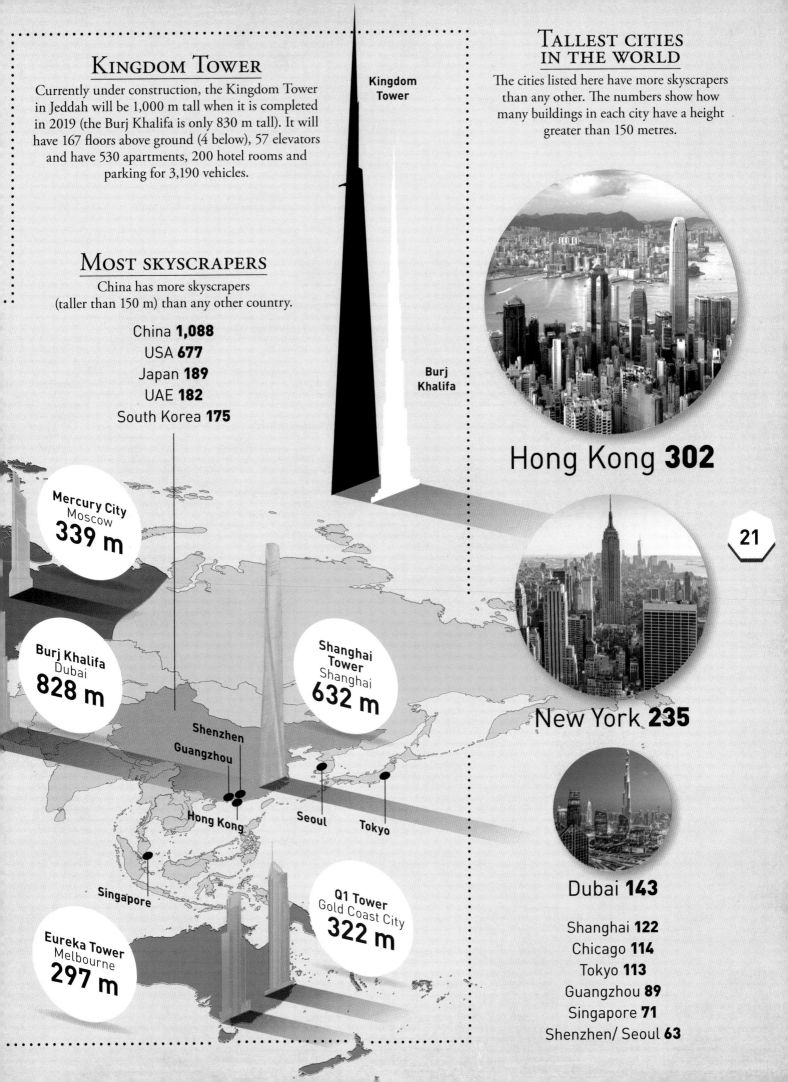

KINGDOM TOWER

Currently under construction, the Kingdom Tower in Jeddah will be 1,000 m tall when it is completed in 2019 (the Burj Khalifa is only 830 m tall). It will have 167 floors above ground (4 below), 57 elevators and have 530 apartments, 200 hotel rooms and parking for 3,190 vehicles.

Kingdom Tower

Burj Khalifa

MOST SKYSCRAPERS

China has more skyscrapers (taller than 150 m) than any other country.

China **1,088**
USA **677**
Japan **189**
UAE **182**
South Korea **175**

TALLEST CITIES IN THE WORLD

The cities listed here have more skyscrapers than any other. The numbers show how many buildings in each city have a height greater than 150 metres.

Hong Kong **302**

21

New York **235**

Dubai **143**

Shanghai **122**
Chicago **114**
Tokyo **113**
Guangzhou **89**
Singapore **71**
Shenzhen/ Seoul **63**

Mercury City
Moscow
339 m

Burj Khalifa
Dubai
828 m

Shanghai Tower
Shanghai
632 m

Shenzhen
Guangzhou

Hong Kong

Seoul

Tokyo

Singapore

Q1 Tower
Gold Coast City
322 m

Eureka Tower
Melbourne
297 m

On the LINE

How people communicate with each other varies from one country to the next. While those living in some countries can use large networks of fixed landlines, other countries do not have this infrastructure and their people tend to use mobile phones.

TELEPHONE ACCESS

This map shows the countries that have the most fixed landlines in total and per head of population, as well as those that have the fewest. It also shows the countries that have the highest numbers of mobile phones – in some cases there are two or three mobile phones for every person.

Germany
50,700,00

Monaco
124

USA
139,000,000

Bermuda
110

Virgin Islands
71

Gabon
215

Cayman Islands
63

Brazil
44,300,000

Falkland Islands
1,980

KEY

Mobile phone subscriptions (per 100 people)

Telephone lines (per 100 people)

Most landlines (total number of main telephone lines)

Least landlines (total number of main telephone lines)

PHONE CALLS

By far the greatest number of phone calls are short-distance calls made between places inside the same country. However, the busiest international phone lines are those running from the USA to Mexico and India. They are most likely used by people living abroad and phoning their friends and families at home.

Mexico

USA

India

LANDLINES VS. MOBILE PHONES

In the year up to June 2014, UK users spent 3 billion fewer minutes on fixed-line phones than they did the year before, a reduction of 12.7 per cent. During the same period, the time spent using mobile phones increased by 2.3 per cent.

12.7%

2.3%

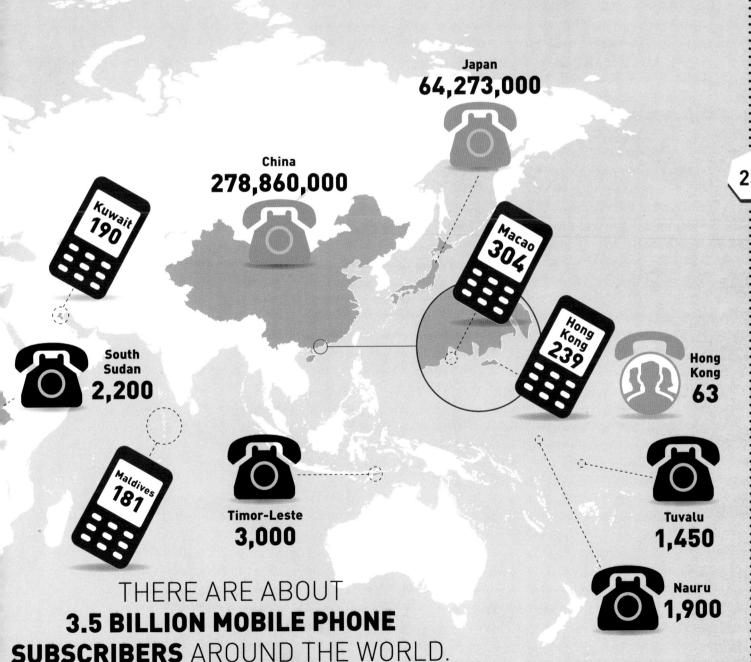

Japan
64,273,000

China
278,860,000

Kuwait
190

Macao
304

Hong Kong
239

Hong Kong
63

South Sudan
2,200

Maldives
181

Timor-Leste
3,000

Tuvalu
1,450

Nauru
1,900

THERE ARE ABOUT
3.5 BILLION MOBILE PHONE
SUBSCRIBERS AROUND THE WORLD.

— Internet —
ACCESS

Since the creation of the internet, how we access the worldwide web has changed. Instead of using a computer on a desk, people are now tending to use smaller devices, such as tablets and smartphones, accessing the internet while they are on the move.

INTERNET USAGE

KEY

 Countries with the most number of personal computer users

 Countries where the greatest percentage of the population uses tablet computers

 Countries with the highest number of internet users per 100 people

 Countries where the greatest percentage of the population owns a smartphone

1.USA
310.6 million

4.USA
20%

2.Bermuda
95.3

PC vs Smartphone vs Tablet

Even though more people will soon be buying a tablet than a computer, computers are still found in more homes around the world. Ultra-portable smartphones, however, are owned by more people than any other internet device and predicted sales for 2015 are close to 2 billion around the world.

20%
Percentage of world population that owns a computer

22%
Percentage of world population that owns a smartphone

6%
Percentage of world population that owns a tablet

In the last five years, sales of smartphones have increased by more than 400 per cent.

Global sales of smartphones

- 2010 – 297m
- 2011 – 472m
- 2012 – 680m
- 2013 – 968m
- 2014 – 1.3bn smartphones

IN 2014, PEOPLE IN THE USA USED MOBILE DEVICES (SMARTPHONES AND TABLETS) TO ACCESS THE INTERNET 55 PER CENT OF THE TIME.

55%

45%

THIS WAS THE FIRST TIME THAT MOBILE DEVICES HAD OVERTAKEN COMPUTERS.

1. Iceland 96.5

5. UK 19%

5. Norway 67.5%

3. Norway 95.1

3. Italy 23%

4. Germany 71.5 million

4. Sweden 94.8

5. Denmark 94.6

2. China 195.1 million

2. South Korea 73.0%

3. Japan 98.1 million

4. Singapore 71.7%

5. India 57 million

Spain 24%

3. Saudi Arabia 72.8%

1. UAE 73.8%

2. Australia – 24%

MONEY

There are about 180 different currencies in use around the world. These are currencies that are approved by governments and exchanged for goods and services. Other ways to pay include cryptocurrencies, such as bitcoin, which are not controlled by national banks or governments.

CURRENCIES AROUND THE WORLD

26

This world map shows the types of currency used around the world. The euro is used in about 20 countries, while the same number use some form of dollar as currency.

TYPES OF CURRENCY

- Dollar/Peso/Real
- Dinar
- Euro
- Rupee/Rupiah
- Franc
- Lira/Pound
- Shilling
- Ruble
- Krona
- Rial/Riyal
- Yen/Yuan/Won
- Other
- No universal currency

ANCIENT CURRENCIES

In the past, people around the world have used a wide range of objects and materials as currency, exchanging them for goods and services. These include salt, animal fur, metal objects and even massive stone wheels.

Kissi money
Used in West Africa around the end of the 19th century, these were T-shaped pieces of iron up to 40 cm long

Squirrel pelts
Used in medieval Russia

Katanga Cross
A copper cross weighing up to 1 kg and used in central Africa until the start of the 20th century

Rai stones
Huge stones measuring more than 3 m across, rai stones were used in Micronesia until the start of the 20th century

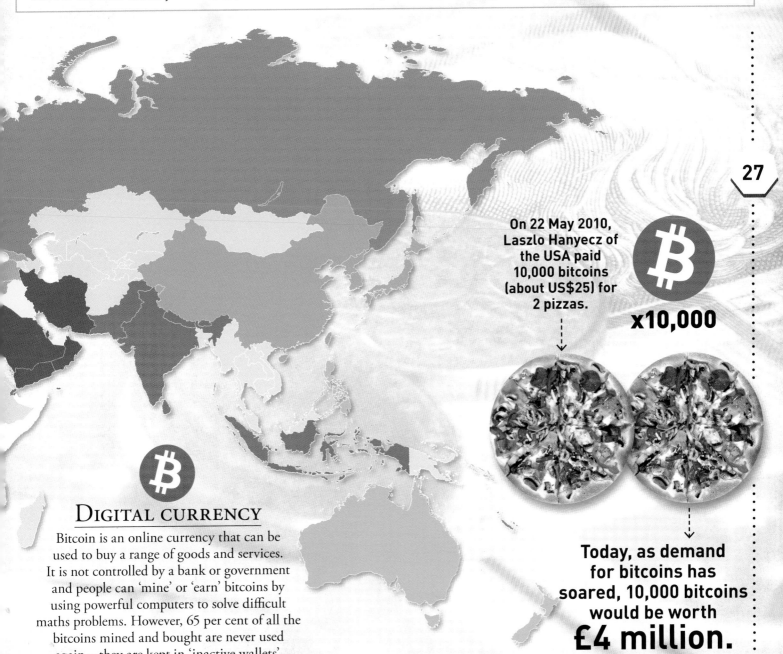

On 22 May 2010, Laszlo Hanyecz of the USA paid 10,000 bitcoins (about US$25) for 2 pizzas.

x10,000

DIGITAL CURRENCY

Bitcoin is an online currency that can be used to buy a range of goods and services. It is not controlled by a bank or government and people can 'mine' or 'earn' bitcoins by using powerful computers to solve difficult maths problems. However, 65 per cent of all the bitcoins mined and bought are never used again – they are kept in 'inactive wallets'.

Today, as demand for bitcoins has soared, 10,000 bitcoins would be worth £4 million.

Space LAUNCHES

As Earth rotates, its surface moves faster at the equator (about 1,675 km/h) than at the poles (where it moves at 0 km/h). For this reason, many of the sites used to launch rockets into space are found between the two tropics either side of the Equator. Rockets can use the speed of Earth's rotation to give them an extra boost.

LAUNCH SITES

This map shows some of the main launch sites around the world. As well as the USA and Russia, places that are actively involved in launching rockets into space include China, India and Europe. There are also a number of private companies that have built and operate their own rockets.

Baikonur, Kazakhstan
The Baikonur Cosmodrome is the world's oldest and biggest space launch site. The first manmade object to orbit Earth, Sputnik 1, and the first human in space, Yuri Gagarin, both blasted off from this facility.

Kennedy Space Center, USA
Based on the east coast of Florida, USA, the Kennedy Space Center has launched every manned NASA flight, including the Apollo missions to the Moon and the Space Shuttle.

Xichang, China

Jiuquan, China

Taiyuan, China

Plesetsk, Russia

Palmachim, Israel

San Marco, Kenya

Wallops, USA

White Sands, USA

Vandenberg, USA

Kodiak, USA

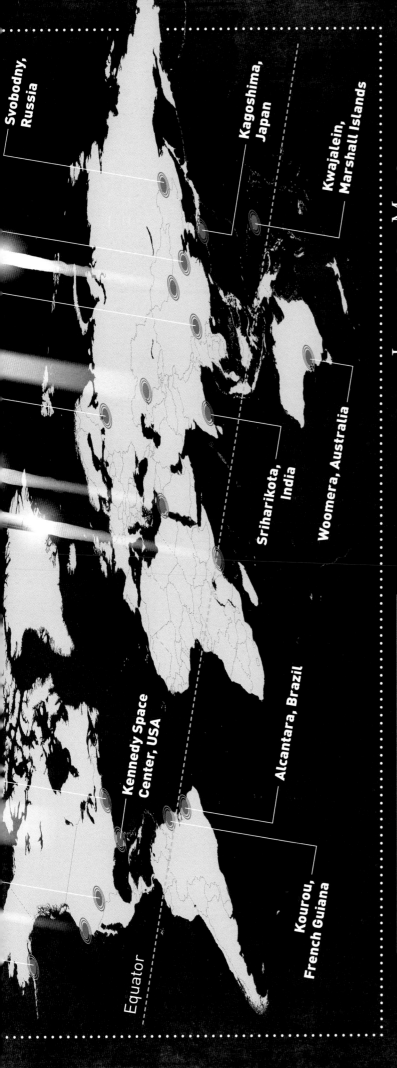

Svobodny, Russia

Kagoshima, Japan

Kwajalein, Marshall Islands

Sriharikota, India

Woomera, Australia

Alcantara, Brazil

Kennedy Space Center, USA

Kourou, French Guiana

Equator

LANDING ON MARS

Seven robot spacecraft have successfully landed on the surface of Mars. Of these, four were static landers, while the other four were rovers and able to travel around the Martian surface, studying the rocks and the atmosphere and sending information and images back to Earth.

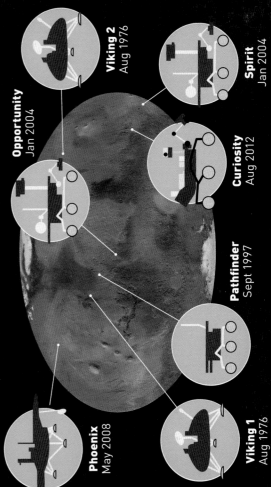

Viking 2 Aug 1976

Spirit Jan 2004

Opportunity Jan 2004

Curiosity Aug 2012

Pathfinder Sept 1997

Phoenix May 2008

Viking 1 Aug 1976

TO THE MOON

To date, only three countries have successfully launched missions to land on the Moon. Of these 20 missions, only six were manned and just 12 people have actually walked on the lunar surface.

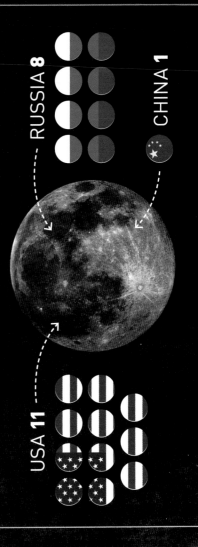

USA 11

RUSSIA 8

CHINA 1

Mapping the WORLD

The maps in this book are two-dimensional representations of our ball-shaped world. Maps allow us to display a huge range of information, including the size of the countries and where people live.

PROJECTIONS

Converting the three-dimensional world into a two-dimensional map can produce different views, called projections. These projections can show different areas of the Earth.

GLOBE
Earth is shaped like a ball, with the landmasses wrapped around it.

CURVED
Some maps show parts of the world as they would appear on this ball.

FLAT
Maps of the whole world show the landmasses laid out flat. The maps in this book use projections like this.

TYPES OF MAP

Different types of map can show different types of information. Physical maps show physical features, such as mountains and rivers, while political maps show countries and cities. Schematic maps show specific types of information, such as routes on an underground train network, and they may not necessarily show things in excatly the right place.

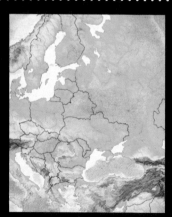

Physical map

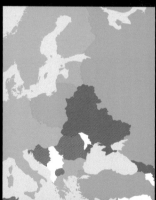

Political map

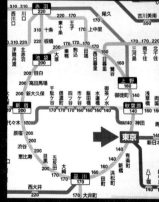

Schematic map

MAP SYMBOLS

Maps use lots of symbols to show information, such as blue lines for rivers and dots for cities. Some of the symbols in this book show the locations of subjects, or the symbols are different sizes to represent different values – the bigger the symbol, the greater the value.

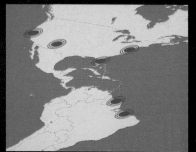

Locator dots

Scaled symbols

CLIMATE CHANGE
The change in Earth's entire climate, and in specific climates around the world.

CRUDE OIL
Oil that has been pumped out of the ground, before it has been refined into other products, such as petroleum.

CRYPTOCURRENCY
A digital form of payment that is used on the internet. It is not controlled by a government or a central bank.

CURRENCY
A method of payment for goods and services, usually in the form of paper notes and metal coins, which is accepted in a country or region. It is usually controlled by a government or a central bank.

EQUATOR
The line that runs horizontally around Earth at its widest part.

EXPORT
To move goods and services out of a country.

HIGH-INCOME COUNTRY
According to the World Bank, this is a country where each person earns more than US$12,736 a year on average.

IMPORT
To move goods and services into a country.

KILOWATT HOUR (KWH)
A unit used to measure energy being transmitted or used. It is equivalent to 1,000 watts per hour and is used to measure electricity usage.

LANDLINE
A fixed connection linking a telephone to a network, rather than a mobile phone connection.

LOW-INCOME COUNTRY
According to the World Bank, this is a country where each person earns less than US$1,045 a year on average.

MEDIUM-INCOME COUNTRY
According to the World Bank, this is a country where each person earns between US$1,046 and US$12,735 a year on average.

PLANTATION
A large piece of land that is being used to grow a single crop, such as the oil palm or sugar cane.

SANITATION
The safe disposal of waste, including human waste, and stopping people from coming into contact with that waste.

SMARTPHONE
A type of mobile phone that can send emails, access the internet and run applications, known as apps.

STOREY
A floor or level in a building. It is usually used when talking about skyscrapers.

TEU
Short for twenty-foot equivalent unit, it is a unit used to describe how much cargo ships can carry. It refers to the large metal containers that are often stacked on the ships' decks.

TROPICS
The regions on Earth that lie on either side of the equator.

WEBSITES

WWW.NATIONALGEOGRAPHIC.COM/KIDS-WORLD-ATLAS/MAPS.HTML
The map section of the National Geographic website where readers can create their own maps and study maps covering different topics.

WWW.MAPSOFWORLD.COM/KIDS/
Website with a comprehensive collection of maps covering a wide range of themes that are aimed at students and available to download and print out.

HTTPS://WWW.CIA.GOV/LIBRARY/PUBLICATIONS/THE-WORLD-FACTBOOK/
The information resource for the Central Intelligence Agency (CIA), this offers detailed facts and figures on a range of topics, such as population and transport, about every single country in the world.

WWW.KIDS-WORLD-TRAVEL-GUIDE.COM
Website with facts and travel tips about a host of countries from around the world.

INDEX

The publisher would like to thank the following for their kind permission to reproduce their photographs:

Key: (t) top; (c) centre; (b) bottom; (l) left; (r) right

Cover front, 1br, 15r isotckphoto.com/Pavlo_K, cover back, 4-5 courtesy of NASA, 1tl, 4-5 light bulbs istockphoto.com/choness, 3c courtesy of NASA, 9br istockphoto.com/Daniel Barnes, 10-11 palm oil fruit istockphoto.com/nop16, 11l istockphoto.com/yotrak, 11br istockphoto.com/kjorgen, 12bl istockphoto.com/Olena

Druzhynina, 14-15 istockphoto.com/David Sucsy, 15bc istockphoto.com, 21tr istockphoto.com/leungchopan, 21cr istockphoto.com/Chris Hepburn, 21br istockphoto.com/Xavier Arnau, 26-27 istockphoto.com/MillefloreImages, 27br istockphoto.com/tavor, 28-29 all courtesy of NASA, 30c istockphoto.com/nicoolay, 30cr istockphoto.com/Manakin

Every attempt has been made to clear copyright. Should there be any inadvertent omission, please apply to the publisher for rectification.

GET THE PICTURE

Welcome to the world of visual learning! Icons, pictograms and infographics present information in a new and appealing way.

9780750278461 — PLANET EARTH
9780750278454 — SPACE
9780750283069 — COUNTRIES
9780750281287 — MACHINES AND VEHICLES
9780750278683 — THE HUMAN BODY
9780750283208 — NATURAL RESOURCES
9780750289856 — THE HUMAN WORLD
9780750283199 — ANIMAL KINGDOM
9780750283229 — SPORT
9780750289863 — THE NATURAL WORLD
9780750283212 — ART AND ENTERTAINMENT
9780750283076 — TECHNOLOGY

9780750297745 — RECORD-BREAKING HUMANS
9780750297653 — RECORD-BREAKING ANIMALS
9780750287470 — RECORD-BREAKING BUILDINGS
9780750297738 — RECORD-BREAKING EARTH & SPACE